YOUR KNOWLEDGE HAS VALUE

- We will publish your bachelor's and
  master's thesis, essays and papers

- Your own eBook and book -
  sold worldwide in all relevant shops

- Earn money with each sale

Upload your text at www.GRIN.com
and publish for free

Bibliographic information published by the German National Library:

The German National Library lists this publication in the National Bibliography; detailed bibliographic data are available on the Internet at http://dnb.dnb.de .

Imprint:

Copyright © 2017 GRIN Verlag, Open Publishing GmbH
Print and binding: Books on Demand GmbH, Norderstedt Germany
ISBN: 9783668518650

This book at GRIN:

http://www.grin.com/en/e-book/374508/design-and-construction-of-a-tensegrity-tower-a-visual-and-statistical

Angelina Aziz

Design and Construction of a Tensegrity Tower. A Visual and Statistical Case Study of a Tensegrity System

GRIN Publishing

TENSEGRITY TOWER

Author

Angelina Aziz

C2 Construction and Dimensioning

S7 Programming and Simulation

Academic Year: 2016/2017

Master of Integrated Architectural Design (MIAD)

Hochschule Ostwestfalen-Lippe
University of Applied Sciences

Part I - Research
Tensegrity structures

Contents

Part II - Construction and Dimensioning C2
Calculation

Part III - Programming and Simulation S7
Visual Study and Conclusion

Part I

Tensegrity structures

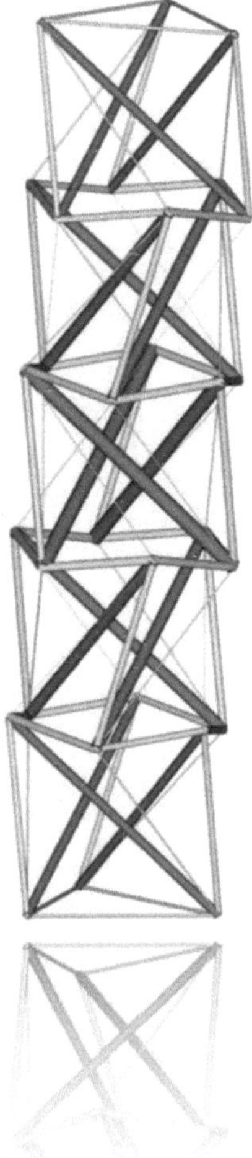

1.1 Definition

The term *tensegrity* is a portmanteau of "tensional integrity" and it was coined by the architect and inventor Richard Buckminster Fuller. Searching on short, sufficient definitions I came across a definition by Valentin Gomez-Jauregui:

> »Tensegrity is a structural principle based on the use of isolated components in compression inside a net of continuous tension, in such a way that the compressed members (usually rods or struts) do not touch each other and the prestressed tensioned members (usually cables or tendons) delineate the system spatially.«

A lot of originally definitions of tensegrity emphasize that the pressure rods do not touch each other.

1.2 Use of tensegrity systems

Until now, tensegrity structures receive less importance in civil engineering, but they are more popular with the visual arts. Tensegrities are of interest in structural design studies because of their lightweight property, aesthetic and modern look. Usually, the structures are built in such a way that struts are connected, which might not be the original definition for tensegrity.

1.3 Open vs. closed tensegrity systems

Open tensegrity systems must, in order to be stable, provide forces to the foundation or to secondary constructions that are beyond the scope of the forces resulting from their own weight and the external loads. Open systems have the advantage that the pressure elements don't have to be used as diagonals, as in closed systems. As a result, shorter pressure sections are possible with open systems, which can be executed with a smaller cross section.

A closed tensegrity system is a self-sufficient array of struts and tendons, arranged in such a way that the struts and tendons enforce an ongoing structural integrity in the overall assemblage. These are termed as "real" tensegrity systems. Closed systems are, regardless of their storage, inherently stable.

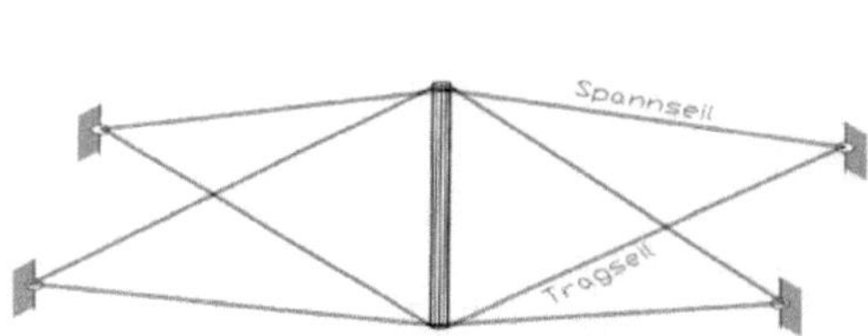

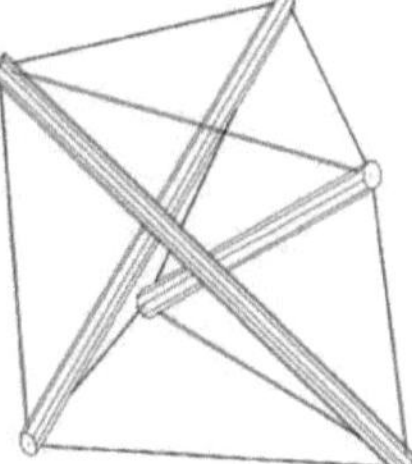

Figure 1.1: Examples for an open tensegrity system (left) and a closed tensegrity systems (right)

1.4 Tensegrity tower

The American artist and pioneer of the tensegrity idea Kenneth Snelson, in contrast to Richard Buckminster Fuller, has dealt more with the artistic design than with the constructional utility of tensegrity structures. An example is his 30 m high Needle-Tower (Figure 1.2) which is neither walkable nor resilient. It can be classified as a pure art object.

In 1965, Snelson was granted the patent for a number of topologies developed by him under the title "Continuous tension, discontinuous compression structures". It also contains the topology of the "Needle Tower" by which Snelson became known worldwide as an artist. For the first time in a large scale such a tower was built in 1968 in Washington D.C on the premises of the "Hirshhorn Museum and Sculpture Garden". The tower is made of aluminum rods and stainless steel ropes and can be lifted by only three persons despite its considerable dimensions (18.3 x 6.2 x 5.4 m).

Figure 1.2: Needle-Tower

1.5 Simplex module

In this project I will examine a two-way column which consists of two simplex modules. One simplex module is composed of three struts and nine tendons (Figure 1.3).

It is important, as in all tensegrity structures, <u>to establish the optimum lengths for the tension members</u> in this simple structure so it will be firm and tightly prestressed. This can only be done by successive adjustments, by trial and error. If the length of one line is changed the tension on all lines are effected.

1.6 Importance of Eurocode

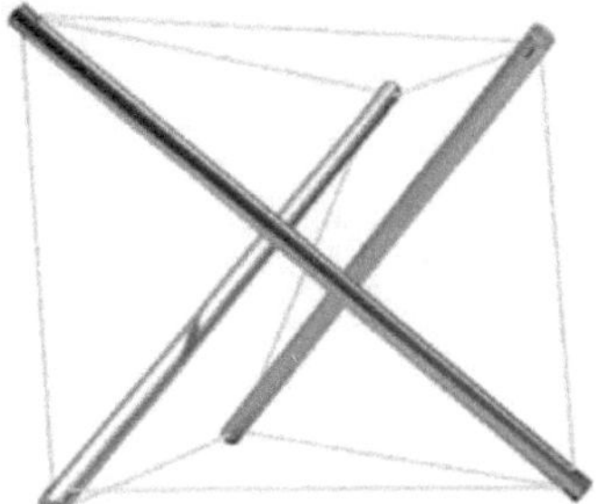

Figure 1.3: Simplex module

»For the EU, the overriding policy objective linked to the Eurocodes is the creation of an Internal Market with the free circulation of products and services while guaranteeing a high level of safety in construction works. Eurocodes will enhance competitiveness of European civil engineering firms, contractors, designers and producers of structural products also in a global context. Indeed, there is a high potential for them being a recognized international reference beyond Europe. Being the result of cross-border cooperation between scientists, experts and national standardization bodies, they can be assumed to correspond to the best and latest state of knowledge, to be universally applicable and to be simply superior to internationally competing standards.«

Retrieved from http://eurocodes.jrc.ec.europa.eu/showpage.php?id=01.

This is why it's necessary to confront all Eurocodes (1-10) for my tensegrity project. With the use of Eurocodes I can calculate a working construction. Nevertheless it's obligatory to check the National Annex because it may contain provisions which complement the Eurocode without inconsistency.

1.6.1 Basis of structural design (EN 1990)

The Eurocode 1990 can be used as a guidance document for design of structures outside the scope of the Eurocodes, so there is the possibility to assess other actions and their combinations.

At first I define the design working life category 1 for my tensegrity tower depending on its planned useful life (Figure 1.4). As a closed tensegrity system my tensegrity tower should be lifted by only two persons, that's why it will be a temporary structure.

Design working life category	Indicative design working life (years)	Examples
1	10	Temporary structures [1]
2	10 to 25	Replaceable structural parts, e.g. gantry girders, bearings
3	15 to 30	Agricultural and similar structures
4	50	Building structures and other common structures
5	100	Monumental building structures, bridges, and other civil engineering structures
(1) Structures or parts of structures that can be dismantled with a view to being re-used should not be considered as temporary.		

Figure 1.4: Indicative design working life

1.6.2 Actions on structures (EN 1991)

Snow load/ Ice load

In this case the calculation of snow load isn't necessary. The tensegrity tower is a sculpture without a roof. But the weight of ice loads plays an important role for the tensegrity system.

Furthermore there are special cases in regard to snow load. Ice loads are generally to be considered with regard to the increased dead weight as well as with regard to their effects on the winds. As a vertical surface load, the maximum ice load can be set at 0.9 kN/m^2 in Germany.

National Annex contains information on ice loads, ice thicknesses, ice densities and ice distributions as well as load case combinations and combination factors for wind and ice effects on towers. The application of Annex C is recommended.

The Eurocode does not give guidance on specialist aspects of snow loading, for example: Effect of wind and ice loading on sag and tension. In cold places there is a ice coating formed on the tendon and also wind pressure nets horizontally on the tendon. Consequently, the ice coating on the tendon increases the total diameter of the tendon and also the weight of the tendon increases. The total weight of the tendon e.g., both the tendon weight and the weight of the ice acts vertically downwards whereas the wind force acts horizontally on the conductor. Therefore, the vector sum of horizontal and vertical forces acting on the conductor gives the total force shown in Figure 1.5.

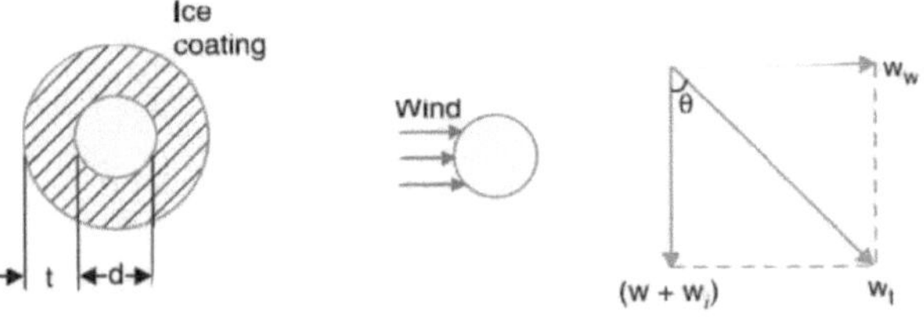

Figure 1.5: Effect of wind and ice loading on sag and tension

Total weight w_t of conductor per unit length is

where

$$w_t = \sqrt{(w + w_i)^2 + (w_w)^2}$$

w = weight of conductor per unit length
= conductor material density $\times$ volume per unit length

w_i = weight of ice per unit length
= density of ice $\times$ volume of ice per unit length
= density of ice $\times \dfrac{\pi}{4} [(d + 2t)^2 - d^2] \times 1$
= density of ice $\times \pi \, t \, (d + t)$

w_w = wind force per unit length
= wind pressure per unit area $\times$ projected area per unit length
= wind pressure $\times [(d + 2t) \times 1]$

Wind pressure

The wind impacts are as a rule described in EN 1991-1-4.
Also the national annex may provide information on the extension of EN 1991-1-4 for towers and masts. The application of the additional rules in Annex B is recommended. Nevertheless it is more important to take a closer look into the German Codes DIN 4131 and DIN 1055-4.

The starting point for the determination of the wind velocity is the map of fundamental basic wind velocity. It is based on a 10-minute mean velocity.

Basic wind velocity: $v_b = c_{dir} \cdot c_{season} \cdot v_{b,0}$

where $v_{b,0}$ = map wind speed (Figure 1.5)

c_{dir} = direction factor

c_{season} = season factor

basic wind velocity for tensegrity tower:

$\underline{v_{b,0}} = 1{,}0 \cdot 1{,}0 \cdot 22{,}5 \text{ m/s} = \underline{\mathbf{22{,}5 \text{ m/s}}}$

Windzone	$v_{b,0}$	$q_{b,0}$
WZ 1	22,5 m/s	0,32 kN/m²
WZ 2	25,0 m/s	0,39 kN/m²
WZ 3	27,5 m/s	0,47 kN/m²
WZ 4	30,0 m/s	0,56 kN/m²

Figure 1.6: Basic wind velocity in Germany

The wind loads should be treated as linear distributed loads on all cable and tube sections and be calculated using the German Code DIN 4131 Anhang A and the draft of the DIN 1055-4 from March 2001. The Codes also regulate by law that wind loads are also to act on cable and tube section covered with ice. That is why it will be necessary to use increased sections, so the wind velocity can be reduced.

Dead weight

The dead weight is usually determined according to EN 1991-1-1.
In addition to that the dead weight of guy ropes must, as a rule, be determined in accordance with EN 1993-1-11.

Preload

As a rule, the required pre-stresses must be set so that the load-bearing structure achieves the required geometrical shape and stress distribution after applying all permanent effects.
The stiffness of my tensegrity tower depends heavily on the preloading. Low preloading would lead to large deflections due to wind and earlier cable drop-out, and possibly to a reduction of bearing capacity of the system. On the other hand, high preloading can also reduce the bearing capacity, e.g. highly compressed tubes might buckle earlier.

Temperature changes

It is of vital importance for the planning of a tensegrity structure to chose the right materials. Temperature changes could have negative effects on the stiffness of the tensegrity structure. In addition to this aspect there should be materials which heat constantly (Figure 1.7). Nevertheless, in the reality where can't be a perfect temperature profile in an outstanding sculpture because of variously solar radiation and shadows.

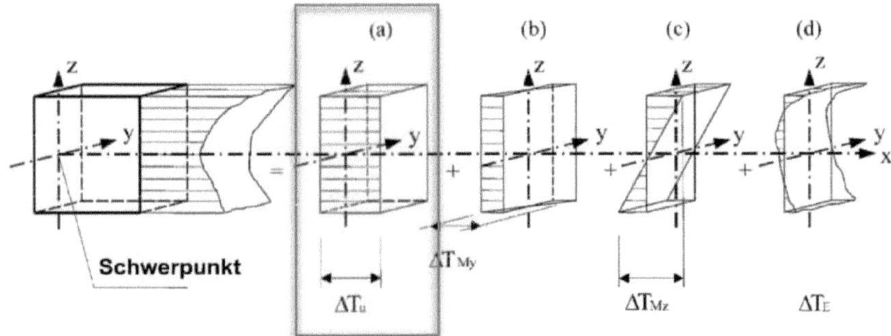

Figure 1.7: Temperature changes

1.6.3 Design of concrete structures (EN 1992)

The loads of a tensegrity system are transferred to the ground through the tensioned cables and the compressed members. The compressed members take on most of the gravity load, dead load, and live load of the structure. When these loads act upon these compressed components, the tensioned cables stretch out an infinitesimally small length to accommodate the increase in stress. Because of this load interaction between the compressed and tensioned components, the only way for the structure to fail is to have the struts buckle or the ties to snap. As all of these forces are pushing und pulling inside the components, the vertical loads continue to transfer down to the ground, where the structure is most likely fixed. This is the reason why many tensegrity-based structures do not require the support of a foundation. Most of the stability of a tensegrity system comes from the geometry of the structure and the materials it is made out of. In addition, the lightweight characteristics of tensegrity structures reduce horizontal wind loads and dead loads drastically.

Figure 1.8: Tensegrity System with foundations

1.6.4 Design of steel structures (EN 1993)

EN 1993-1-1 contains rules for the design, calculation and dimensioning of steel structures with sheet thicknesses t ≥ 3 mm. In addition, application rules for building construction are specified.

Section 5 relates to the calculation of the load bearing capacity of bar support structures, and section 6 contains detailed rules for the design of cross-sections and components in the limit of load-bearing capacity. For the tensegrity sculpture it is possible to take steel structure with a smaller sheet thickness.

1.6.5 Design of composite steel and concrete structures (EN 1994)

This Eurocode isn't useful for the tensegrity tower because this project can't deal with composite steel.

1.6.6 Design of timber structures (EN 1995)

EN 1995 addresses the requirements for load-bearing capacity, usability, durability and fire resistance of wooden structures. Other requirements, such as heat and noise protection, are not covered. In regard to the tensegrity tower the mechanical properties of timber are moisture depended. A change of moisture content from 12% to 20 % leads to significant reduction (Figure 1.9).

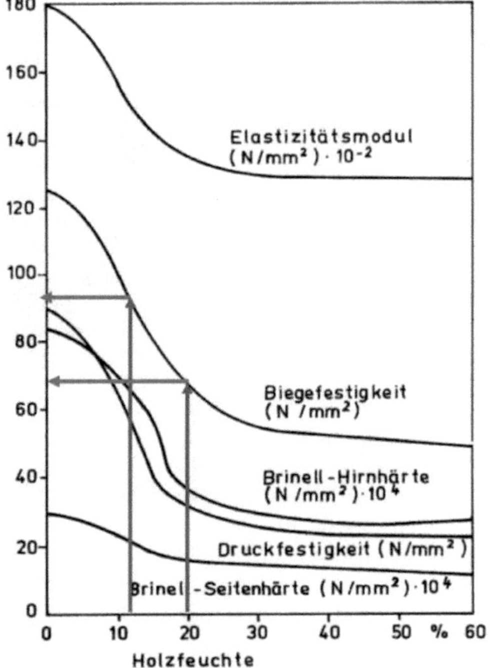

Figure 1.9: Wood moisture

1.6.7 Design of masonry structures (EN 1996)

This Eurocode isn't useful for the tensegrity tower because this project can't deal with masonry structures.

1.6.8 Geotechnical design (EN 1997)

The Eurocode 1997 has useful hints for tensegrity structures with foundations. In the planning of supporting structures, the temporal and spatial effects of unusual temperature changes must be taken into account. In order to prevent the formation of ice lenses in the ground behind the supporting structure, special precautions such as, for example, the selection of suitable filling material, drainage or insulation must be taken.

1.6.9 Design of structures for earthquake resistance (EN 1998)

The stiffness of the foundations shall be adequate for transmitting the actions received from the superstructure to the ground as uniformly as possible. With the exception of bridges, only one foundation type should in general be used for the same structure, unless the latter consists of dynamically independent units.

1.6.10 Design of aluminium structures (EN 1999)

The Eurocode EN 1999 shows several characteristic attributes of aluminium structures. In the selection of material for the rods, it could be helpful to take a closer look in to this table (Figure 1.10).

Tabelle 3.2a — Charakteristische Werte der 0,2 %-Dehngrenze f_0 und der Zugfestigkeit f_u (ungeschweißt und für WEZ), Mindestwerte der Bruchdehnung A, Abminderungsfaktoren $\rho_{o,haz}$ und $\rho_{u,haz}$ in der WEZ, Knickklasse und Exponent n_p für Aluminiumknetlegierungen — Bleche, Bänder und Platten

Legierung EN-AW	Zustand[1]	Dicke t [1]	f_o [1]	f_u	A_{50} [1, 6]	$f_{o,haz}$ [2]	$f_{u,haz}$ [2]	WEZ-Faktor[2]		BC [4]	n_p [1, 5]
		mm	N/mm²		%	N/mm²		$\rho_{o,haz}$ [1]	$\rho_{u,haz}$		
3004	H14 | H24/H34	≤ 6 | 3	180 | 170	220	1 | 3	75	155	0,42 | 0,44	0,70	B	23 | 18
	H16 | H26/H36	≤ 4 | 3	200 | 190	240	1 | 3			0,38 | 0,39	0,65	B	25 | 20
3005	H14 | H24	≤ 6 | 3	150 | 130	170	1 | 4	56	115	0,37 | 0,43	0,68	B	38 | 18
	H16 | H26	≤ 4 | 3	175 | 160	195	1 | 3			0,32 | 0,35	0,59	B	43 | 24
3103	H14 | H24	≤ 25 | 12,5	120 | 110	140	2 | 4	44	90	0,37 | 0,40	0,64	B	31 | 20
	H16 | H26	≤ 4	145 | 135	160	1 | 2			0,30 | 0,33	0,56	B	48 | 28
5005/ 5005A	O/H111	≤ 50	35	100	15	35	100	1	1	B	5
	H12 | H22/H32	≤ 12,5	95 | 80	125	2 | 4	44	100	0,46 | 0,55	0,80	B	18 | 11
	H14 | H24/H34	≤ 12,5	120 | 110	145	2 | 3			0,37 | 0,40	0,69	B	25 | 17
5052	H12 | H22/H32	≤ 40	160 | 130	210	4 | 5	80	170	0,50 | 0,62	0,81	B	17 | 10
	H14 | H24/H34	≤ 25	180 | 150	230	3 | 4			0,44 | 0,53	0,74	B	19 | 11
5049	O / H111	≤ 100	80	190	12	80	190	1	1	B	6
	H14 | H24/H34	≤ 25	190 | 160	240	3 | 6	100	190	0,53 | 0,63	0,79	B	20 | 12
5454	O/H111	≤ 80	85	215	12	85	215	1	1	B	5
	H14|H24/H34	≤ 25	220 | 200	270	2 | 4	105	215	0,48 | 0,53	0,80	B	22 | 15
5754	O/H111	≤ 100	80	190	12	80	190	1	1	B	6
	H14|H24/H34	≤ 25	190 | 160	240	3 | 6	100	190	0,53 | 0,63	0,79	B	20 | 12

Figure 1.10: Table with characteristic attributes of aluminium structures

Part II - Construction and Dimensioning C2

Calculation

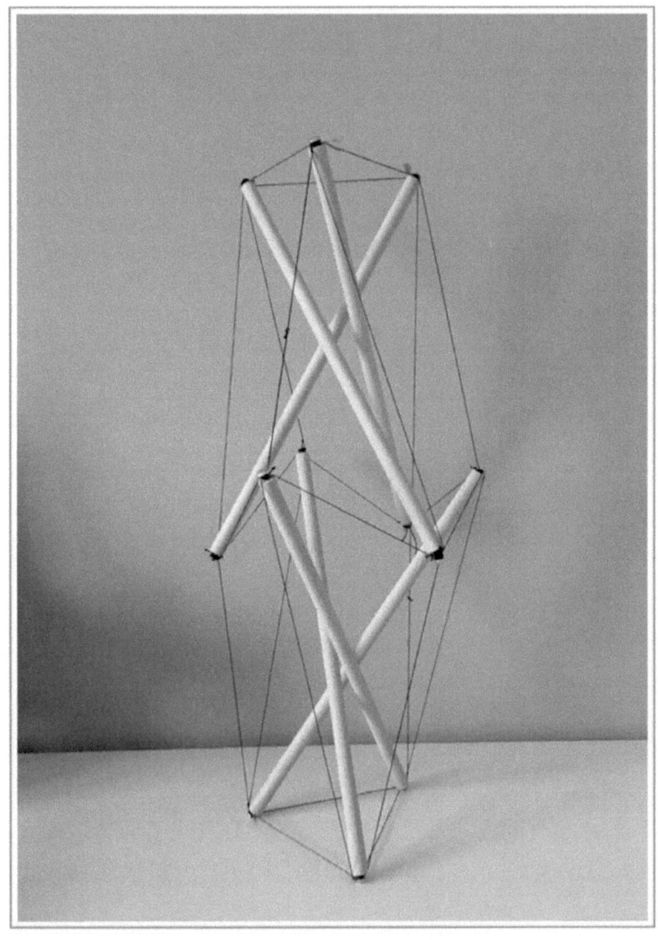

2.1 Basis of calculation

2.1.1 Terms

Before I start to calculate my tensegrity tower, there is the need to clarify all terms in a simplex module (Figure 2.1). The ropes each form a triangular basis polygon as a base and as a top surface. These two triangles are rotated relative to each other about a twisting angle. In the plan view, the uppermost connection of a corner of the upper polygon to a corner of the lower polygon is performed as a rope and the longest connection as a rod. The two triangles are equilateral, but can have different lateral lengths. The horizontal ropes forming the polygons of the base and top surfaces are referred to as polygons, and the vertical ropes connecting the triangles vertically at the outer edge of the module are referred to as vertical ropes. Furthermore, the concept of the longitudinal axis, which is to pass through the centers of the two triangles, is frequently used. The vertical distance between the two basic polygons is the height h_{ds} of the simplex module.

I will define the lower simplex module as simplex module (A) and the upper simplex module as simplex module (B).

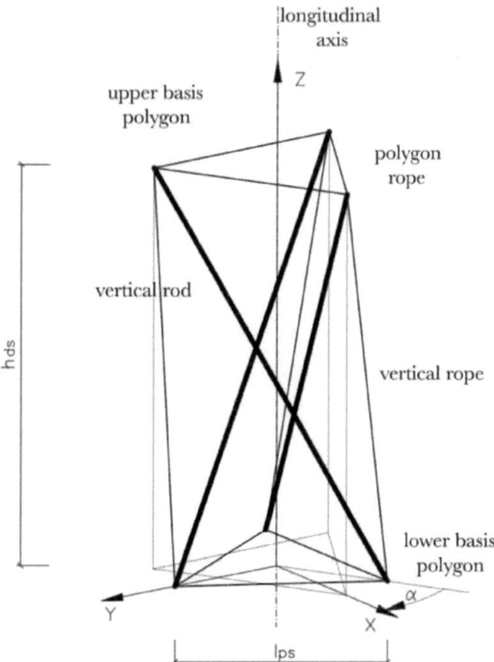

Figure 2.1: Terms and geometrical parameters of a tensegrity simplex module

2.2 Static methods

2.2.1 Analytical solution

It is also possible to look at the statics by looking at the tensegrity as a simple framework. To do this, the static determinateness of the object (Figure 2.2) should be checked.

n = r + s − 3k

r : Number of reactions in the bearing. For tensegrities, r = 0, because there are no bearings
s : Number of rods and ropes
k : Number of nodes

for n < 0 : statically undetermined, displaceable
for n = 0 : statically determined, at rest
for n > 0 : statically overstated, at rest (safer than n = 0)

n = 0 + 30 − 36 = - 6

=> statically undetermined, displaceable

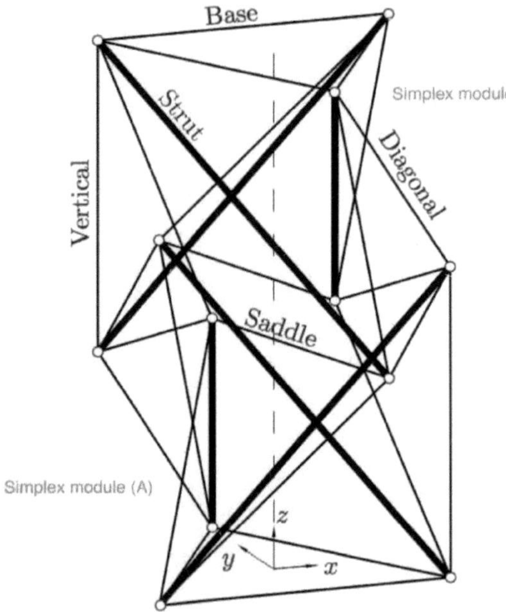

Figure 2.2: Two simplex modules with diagonal ropes

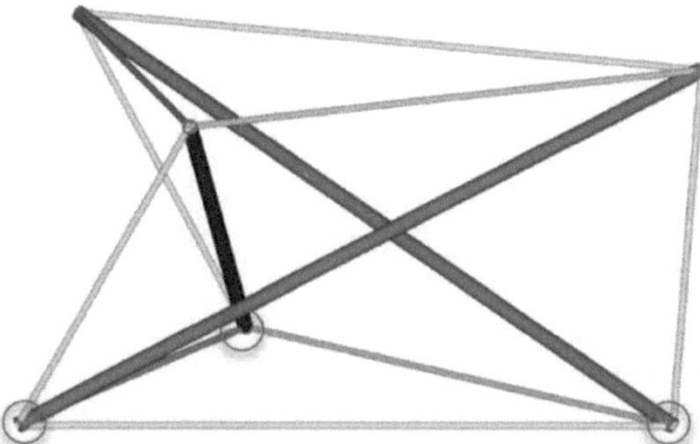

Figure 2.3: Two simplex modules with diagonal ropes

But if there would be reactions in the bearing of one simplex module ($r = 6$) like you can see in Figure 2.3, <u>the system would be statically determined</u>:

n = 6 + 12 − 12 = <u>0</u>

The static indeterminacy which has been calculated in Figure 2.2 has the effect that tensegrities are generally extremely unstable to external stress without bearing and can be easily deformed. Tensegrities greatly depend on how strong the pretension of the ropes is, that means under how much pull it was set up. Depending on this, it can withstand considerable external forces. The construct shifts into a new equilibrium position.

If, however, it is taut enough, the critical point can be retarded until the whole system becomes unstable.

From the fundamentals of statics, we know that we can only compute statically determined systems relatively easily. The rod and rope forces can therefore be calculated directly from the equilibrium conditions.

In 1978, Calladine in (Calladine1978) published the extension of Maxwell's law. With this, it is possible to record all the special cases.

$$3n - e - c = m - s$$

n : Number of nodes
e : Number of elements
c : Number of reaction in the bearing
m: Number of possible motions
s : Number of preload states
c : Number of reaction in the bearing

Structures with m > 0 are defined as kinematically indeterminate and structures with s > 0 are designated as unstressed (Pellegrino1986). Pellegrino subdivided (Pellegrino1993) the rod supports on the basis of the values m and s into four types. Table (2.4) shows this classification, supplemented by the usefulness of the individual types as a framework.

Type	m and s	Characteristics
1	s = 0	- structure is stable - can not be preloaded
1	m = 0	useable
2	s = 0	- structure is not stable - can not be preloaded
2	m > 0	not useable
3	s > 0	- structure is stable - can be preloaded
3	m = 0	useable
4	s > 0	- structure is displaceable - can be preloaded
4	m > 0	conditionally useable

Figure 2.4: Table with characteristics of individual framework types

The two simplex modules (A) and (B) are arranged in such a way that the ropes of the upper basis polygon floor are in the middle of the upper polygon rope of the lower simplex module (A). That results that the intermediate level contains a hexagon. Additional diagonal ropes connect the floors to each other. The ropes extend from the lower nodes of the upper   module to the lower nodes of the lower module and from the upper nodes of the upper   module to the upper nodes to the lower module. Thus, in each basis polygon, there are three ropes, and 12 additional diagonal ropes.

For a tower with n_e floors, the number of ropes is calculated as:

$$e_s = 3(5n_e - 2) = \underline{24\ \text{ropes}}$$

Taking three rods per simplex module into account, the number of elements e to:

$$e = 3(6n_e - 2) = \underline{30\ \text{elements}}$$

The number of nodes is:

$$n = 6n_e = \underline{12\ \text{nodes}}$$

My tensegrity structure with two floors has $e = 30$ elements according to equations and $n = 12$ nodes. Thus, Mawell's law ($3n - e - c = m - s$) is precisely met. This means that it can be preloaded in its general geometry.

=> m - s = 6

If my tensegrity can be prestressed and thus allow the use of ropes and rods, it is necessary to find a specific geometry of the tensegrity state. In the case of the single three-step module, the crucial geometry parameter, which differs between general and special geometry, is the twist angle of the base polygons.

In order to be able to pretension tensegrity of this type, it is necessary to achieve the tensegrity state in each individual module. Therefore the twist angle of the base polygons, the height h_{ds}, and the number of floors play an important role.

2.3 Kinematic methods

2.3.1 Analytical solution

All kinematic methods have a common search for an extreme value of the rod or rope length. The maximum rod length is determined for a constant length of the ropes and the minimum length for a constant length of the rod.

For symmetric tensegrity modules, e.g. a simplex module, it is possible to find the form with the help of a closed analytical solution. The distance between the two base polygons h_{ds} is calculated as a function of the rod length l_d, the radius r and the twist angle α of the base polygons:

$$h_{ds} = \sqrt{l_d^2 - 4r^2 \sin\left(\frac{\pi}{3} + \frac{\alpha}{2}\right)^2} \qquad\qquad (1.1)$$

Example 1a):

For a defined rod length $l_d = 2$ and a fixed radius r = 1:

$$h_{ds} = 2\sqrt{1 - \sin\left(\frac{\pi}{3} + \frac{\alpha}{2}\right)^2} \quad = 1{,}92 \qquad\qquad (1.2)$$

The length of the vertical ropes l_{vs} is then:

$$l_{vs} = 2\sqrt{\sin\left(\frac{\alpha}{2}\right)^2 + 1 - \sin\left(\frac{\pi}{3} + \frac{\alpha}{2}\right)^2} \quad = 1{,}99 \qquad\qquad (1.3)$$

For a form finding, the extreme values, more precisely the <u>minimum values</u>, are now searched for:

$$\frac{dl_{vs}}{d\alpha} = \frac{\sin\left(\frac{\alpha}{2}\right)\cos\left(\frac{\alpha}{2}\right) - \sin\left(\frac{\pi}{3} + \frac{\alpha}{2}\right)\cos\left(\frac{\pi}{3} + \frac{\alpha}{2}\right)}{\sqrt{\sin\left(\frac{\alpha}{2}\right)^2 + 1 - \sin\left(\frac{\pi}{3} + \frac{\alpha}{2}\right)^2}} \qquad\qquad (1.4)$$

Within a full circle, this function has three zeros, with only one leading to a minimum value of equation (1.3). This zero exists at:

$$\alpha = \frac{\pi}{6} \quad = 0{,}52 \implies \text{convert into degrees (30°)} \qquad\qquad (1.5)$$

In the interval of 2π, this zero point is repeated once. It should be mentioned that a twist angle of $\alpha = -\frac{\pi}{6}$ also leads to a tensegrity state. The three-bar module is the simplest, spacious tensegrity structure with two parallel polygonal top surfaces. Denoting the number of corners of the polygons with n, Equation (1.5) according to (Motro2003) represents a special case of the following equation which holds for any basic polygons:

$$\alpha = \frac{\pi(n-2)}{2n} = 0{,}52 \tag{1.6}$$

The simple module is triple-rotationally symmetric and it is possible to look for an analytical solution with one-third of the elements. However, this method is not applicable to non-symmetric structures with multiple variables.

2.3.2 Definition of the overlap h_o

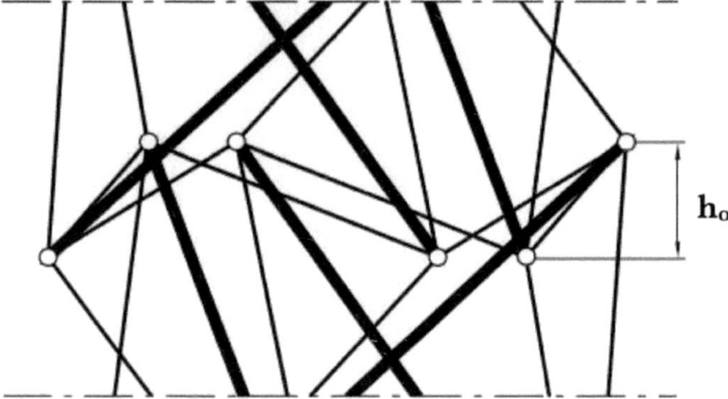

Here, a (= 1) is the side length of the equilateral triangles at the top and bottom of the tower, and l the length of each rod. A particular symmetric configuration, in which all the nodes lie on the surface of a cylinder, is defined by the following relationship between δ and α:

$$\delta = \arcsin\left[\frac{2a}{l\sqrt{3}}\sin\left(\alpha + \frac{\pi}{3}\right)\right]$$

The definition of the overlap h_o:

$$h_o = \begin{cases} \dfrac{1}{2\tan\delta\cos\left(\alpha + \dfrac{\pi}{6}\right)}\left(\sqrt{\dfrac{a^2}{3} - 3l^2\sin^2\delta\cos^2\left(\alpha + \dfrac{\pi}{6}\right)} \right. \\ \left. -\dfrac{a}{\sqrt{3}} + l\sin\delta\cos\left(\alpha + \dfrac{\pi}{6}\right)\right) \\ \dfrac{l\cos\delta}{2} \end{cases}$$

2.3.3 Determination of the additional twisting angle

$$\varphi \;=\; 180° \cdot \left(\frac{1}{2} - \frac{i}{n}\right) \;,\; \text{für } 1 \leq i < n$$

$$l_z = \sqrt{l_D^2 + 2 \cdot r_1 \cdot r_2 \cdot [\cos(i \cdot \beta + \varphi) - \cos \varphi]}$$

$$h = \sqrt{l_D^2 + 2 \cdot r_1 \cdot r_2 \cdot \cos(i \cdot \beta + \varphi)}$$

$$a_1 = 2 \cdot r_1 \cdot \sin\left(\frac{\beta}{2}\right) \qquad a_2 = 2 \cdot r_2 \cdot \sin\left(\frac{\beta}{2}\right) \qquad \beta = \frac{360°}{n}$$

Figure 2.4: Formula to calculate the additional twisting angle

Example 1b):

φ = additional twisting angle = 30°
β = angle between lower points or upper points adjacent rods = 120°
i = number of possible starting positions of rods = 1
n = number of rods; number of corners of a basis polygon = 3
l_z = length of vertical ropes = 35,41 cm
h = height of a simplex module = 37,77 cm
l_D = length of a rod = 40 cm
r_1 = radius of lower basis polygon = 10 cm
r_2 = radius of upper basis polygon = 10 cm
a_1 = length of ropes between rods in the lower basis polygon = 17,32 cm
a_2 = length of ropes between rods in the upper basis polygon = 17,32 cm

2.4 Load calculation

Example 2):

We have an outstanding tensegrity sculpture with following information:

- tower contains 2 simplex module
- 6 pvc pipes (d = 50 mm; 1 m = 0,9 kg)
- 24 stainless steel wire ropes (d = 5 mm; 1 m = 0,1 kg)
- height of tower: 1,85 m
- width of tower: 0,5 m

Dead load:

space load of pvc: 15 kN/m^3

volume of a pvc pipe: 0,002 m^2 x 1 m = 0,002 m^3

dead load: 0,002 m^3 x 15 kN/m^3 = <u>0,03 kN/m^3</u>

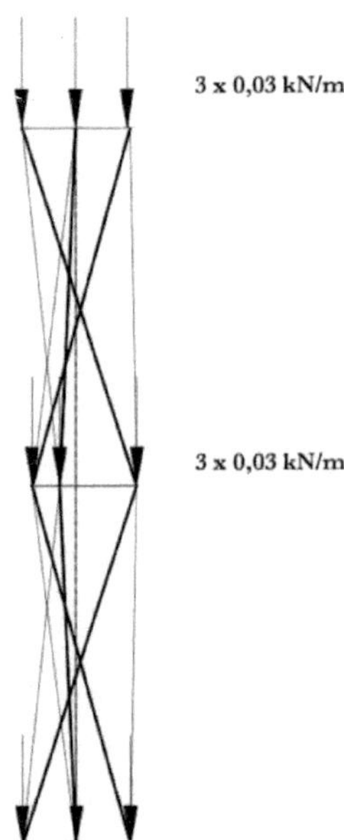

Total load:

Dead weight + live load:

<u>0,03 kN/m^3 x 6 + 1 kN/m^3 = 1,18 kN/m^3</u>

This simple tensegrity tower weighs a total of 7,8 kg. The total load of this structure is 1,18 kN/m^3 and it contains a specified load (1 kN/m^3). These loads come in many different forms, such as people, equipment, vehicles, wind, rain, snow, earthquakes, the building materials themselves, etc. Specified loads also known as characteristic loads in many cases.

Wind load:

$W = c_f \times q \times A$

$\underline{= 1{,}3 \times 0{,}5 \times (0{,}5 \times 1{,}85 \times 0{,}4) = 0{,}24 \text{ kN}}$

c_f : aerodynamic coefficient (1,3)
A : reference surface (here i take 40 % of the reference surface of the tensegrity tower)
q : dynamic pressure (0,5)

dynamic pressure q:

height above ground m	0 - 8	>8 - 20	>20 - 100	>100
q [KN/m²]	0,5	0,8	1,1	1,3

some examples for wind load calculation:

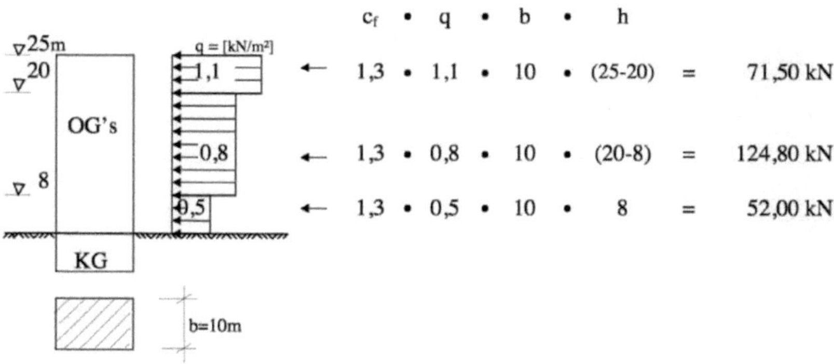

2.5 Tensegrity modell

2.5.1 Prototyping

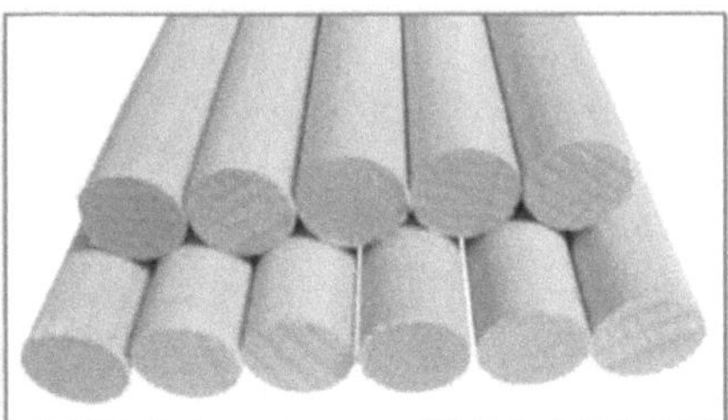

wooden round rods

material	diameter	length	advantages	disadvantages
wooden round rods	14 mm	20 cm	- good for prototyping - cheap material	- wood moisture - movement of wood

rubber cord

material	diameter	length	advantages	disadvantages
rubber cord	1 mm	10 m	- good for preloading - cheap material	

polyester rope

material	diameter	length	advantages	disadvantages
polyester rope	3 mm	10 m		- to loose for tensegrity

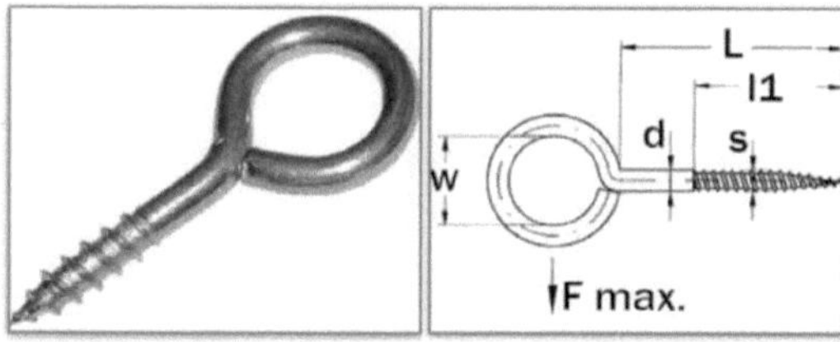

ring bolt

material	diameter	length	advantages	disadvantages
ring bolt	d = 2,8 mm s = 3,6 mm l = 16 mm w = 6 mm l1 = 10 mm	16 mm	- easy to screw - big hole for knots	- the right direction of the ring bolt is important

2.5.2 Final modell

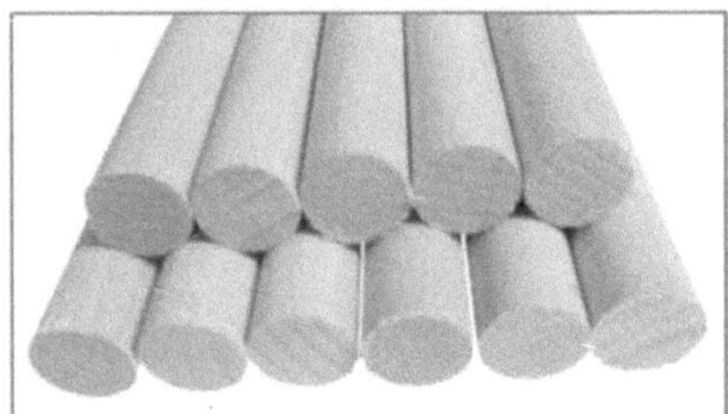

wooden round rods

material	diameter	length	advantages	disadvantages
wooden round rods	14 mm	40 cm	- good for prototyping - cheap material	- wood moisture - movement of wood

mason's lacing cord

material	diameter	length	advantages	disadvantages
mason's lacing cord	1,3 mm	100 m	- good for prototyping - cheap material	- bad for preloading

upholstery nails

material	diameter	length	advantages	disadvantages
upholstery nails	10 mm	12,5 mm	- easy to screw	

rubber cord

material	diameter	length	advantages	disadvantages
rubber cord	1 mm	10 m	- good for preloading - cheap material	

2.5.3 Modell photos

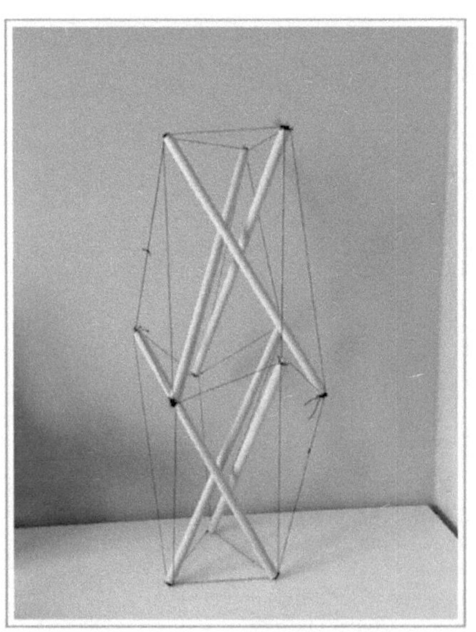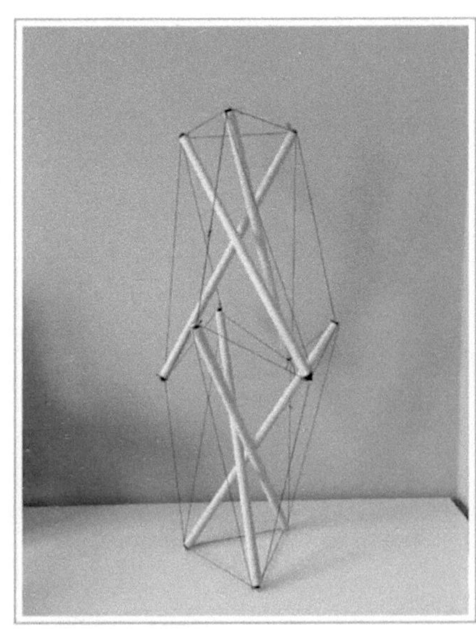

Part III -
Programming and Simulation S7

3.1 Calculation with python

For prototyping or creating one simplex module it is really helpful to take python for calculation. I use the formula for additional twisting angle (2.3.3). As a result I get the length of the rods and the height of a simplex module depending on some given information (r1; i; n).

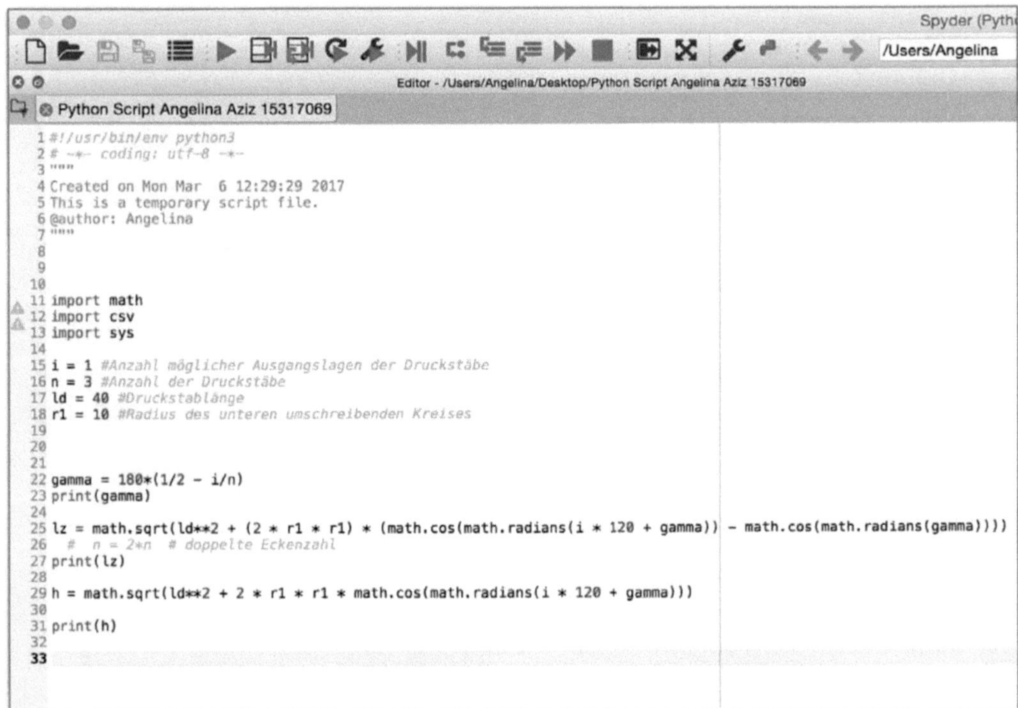

```python
#!/usr/bin/env python3
# -*- coding: utf-8 -*-
"""
Created on Mon Mar  6 12:29:29 2017
This is a temporary script file.
@author: Angelina
"""



import math
import csv
import sys

i = 1 #Anzahl möglicher Ausgangslagen der Druckstäbe
n = 3 #Anzahl der Druckstäbe
ld = 40 #Druckstablänge
r1 = 10 #Radius des unteren umschreibenden Kreises



gamma = 180*(1/2 - i/n)
print(gamma)

lz = math.sqrt(ld**2 + (2 * r1 * r1) * (math.cos(math.radians(i * 120 + gamma)) - math.cos(math.radians(gamma))))
  #  n = 2*n  # doppelte Eckenzahl
print(lz)

h = math.sqrt(ld**2 + 2 * r1 * r1 * math.cos(math.radians(i * 120 + gamma)))

print(h)
```

At first I had some problems with my solutions because it's important to use the radiant in cosine. After correcting my errors, the script is working.

3.2 Visual study

In order to find the perfect relation between the width (D) and the height (H) of my tower, I did a visual study.

With the help of grasshopper, my tensegrity tower can be modified in the graphical view or by editing some parameters (Figure 3.2).
Based on the Golden Section (ration phi = 1 : 1.6), I found the perfect sculpture with the best aesthetic sensitivity (Figure 3.4).

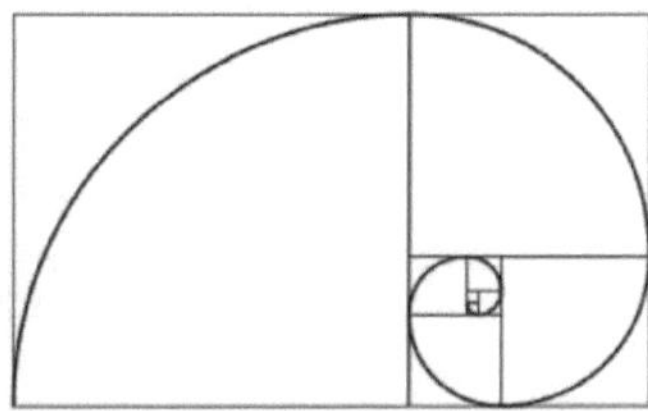

Figure 3.1: Golden Ratio

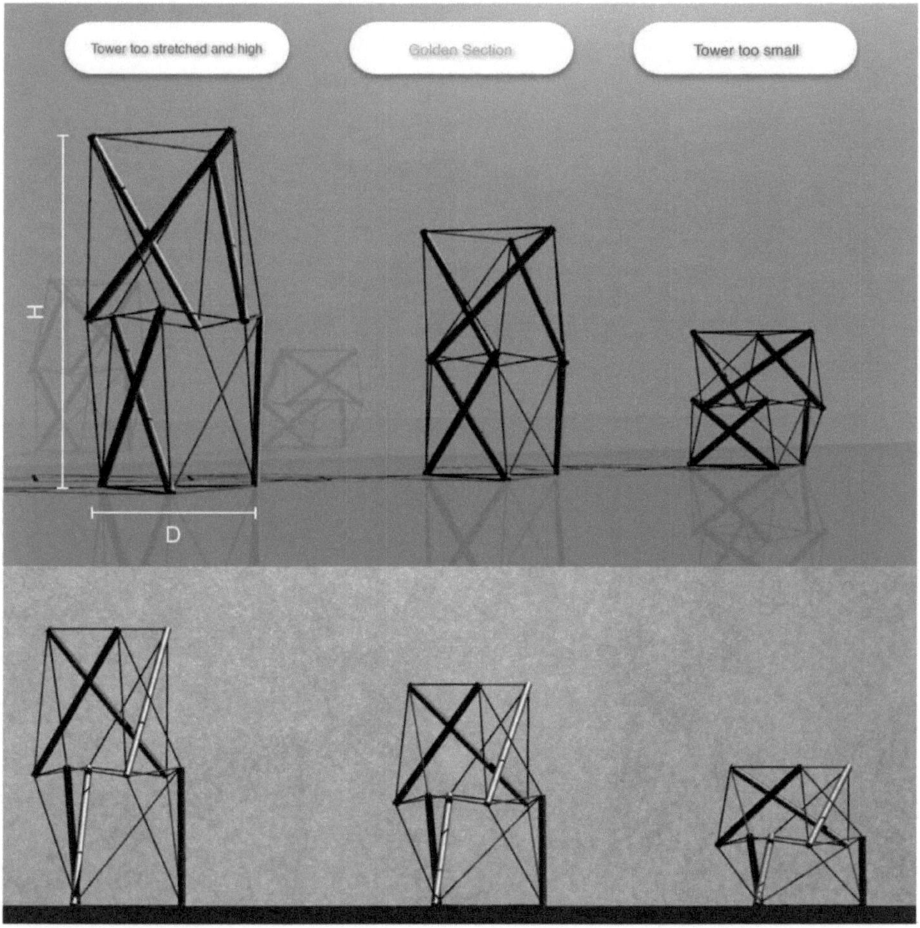

Figure 3.2: Visual study

Figure 3.3: Visual study (here: perspective)

Figure 3.4: Tensegrity sculpture performs the golden ratio

3.3 Grasshopper

For a complete visual simulation, CAD models have a high importance. I used rhinoceros and grasshopper for modelling. Grasshopper is a graphical algorithm editor tightly integrated with rhino's 3-D modelling tools. It is primarily used to build generative algorithms, such as for generative art. Advanced uses of grasshopper include parametric modelling for structural engineering, parametric modelling for architecture and fabrication.

In grasshopper I can regulate :

- height of each simplex module
- the radius of lower and upper basis polygons
- height of overlap
- additional twisting angle
- thickness of rods
- thickness of ropes

After form finding I can bake my two simplex module tower (Figure 3.5).

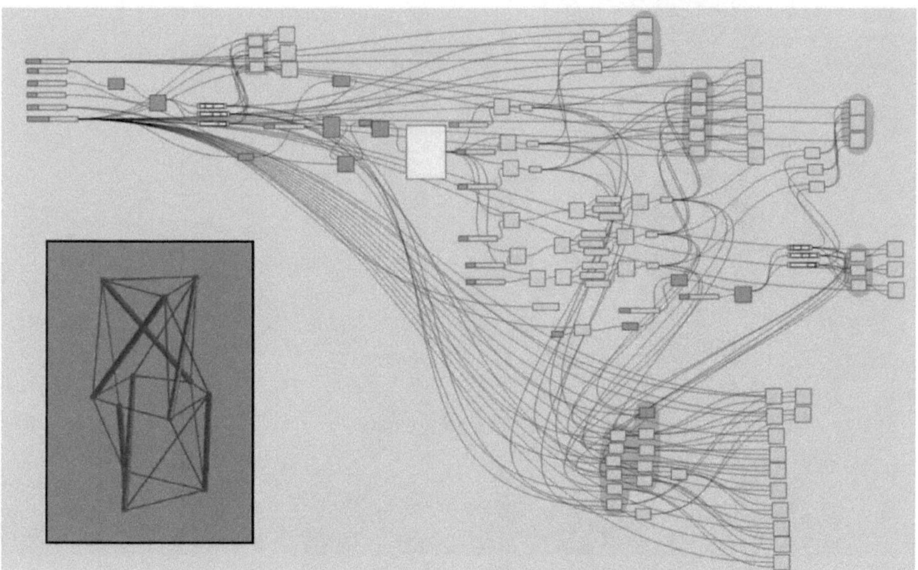

Figure 3.5: Simulation with grasshopper

3.4 Conclusion

All in all tensegrity structures are structures based on the combination of some simple design pattern. The loading members are only in pure compression or in pure tension. That means the structure will only fail if the cables yield or the rods buckle. Also tensional prestress allows cables to be rigid in tension. In regard to these patterns, no structural member experiences a bending moment. This can produce exceptionally rigid structures for their mass and for the cross section of the components.

In my opinion "tiny tensegrity towers" have a modern and extraordinary look. They would go well with the Computational Design program in Hochschule Ostwestfalen-Lippe. In order to give the campus a new computational look, I think it could be good to place some tensegrity towers around the campus. The towers are good meeting points for students. In darkness, an interesting light effect could be obtained by the rods. Also the lightweight for a simple transport makes the tensegrity tower ideal for fairs and events. It can be used as a design object or luminaire element. It is definitely an eye catcher.

Figure 3.6: Visualisation (place: Campus of Hochschule Ostwestfalen-Lippe Detmold)

4.1 Source directory

Websites:

- First Page: The mathematical version of tension elements within a floating tension surface: https://cerebrovortex.com/2013/08/19/silk-island/

- https://en.wikipedia.org/wiki/Tensegrity

- http://www.sciencedirect.com/science/article/pii/S1877705815013338

- http://www.tensegridad.es/Publications/MSc_Thesis-Tensegrity_Structures_and_their_Application_to_Architecture_by_GOMEZ-JAUREGUI.pdf

- https://core.ac.uk/download/pdf/18619456.pdf

- http://civilenggseminar.blogspot.de/2011/09/1.html

Dissertations:

- Dipl.-Ing. Christian Wolkowicz, Ein Beitrag zur Evolution des Tensegrity - Konzeptes
- Gunnar Tibert, Deployable Tensegrity Structures for Space Applications Gunnar
- Jingyao Zhang, Structural Morphology and stability of tensegrity structures